Anonymous

Silversmith's Art and Ecclesiastical Metal Work at its Home

Benziger Brothers' Factory

Anonymous

Silversmith's Art and Ecclesiastical Metal Work at its Home Benziger Brothers' Factory

ISBN/EAN: 9783337307172

Printed in Europe, USA, Canada, Australia, Japan

Cover: Foto ©berggeist007 / pixelio.de

More available books at **www.hansebooks.com**

SILVERSMITH'S ART

AND

ECCLESIASTICAL METAL WORK

AT ITS HOME

BENZIGER BROTHERS' FACTORY

DE KALB AVE. & ROCKWELL PLACE, BROOKLYN, N. Y.

Pontifical Institute of Christian Art

BENZIGER BROTHERS' SALESROOMS

NEW YORK, 36 & 38 BARCLAY STREET

CINCINNATI, 143 MAIN STREET

CHICAGO, 178 MONROE STREET

ST ELIGIUS.   PATRON OF SILVERSMITHS.

It was in 1864 that we took charge of a small silver-smiths' shop on one floor of a building in Fulton Street. New York, and began on our own account the manufacture of Church goods, for these goods had been made for us in this country as far back as 1853.

Ours was not a very extensive establishment in those days, being confined to the manufacture of gold and silver goods on a small scale, but it was the nucleus of our present factory.

The new branch gave fresh impetus to our business, and we applied ourselves to it with all our energy, and when, in 1871, we found it had grown too large for its quarters we rented the buildings 43 and 45 Dey Street. Twenty years later we still further enlarged the business by beginning the manu-facture of candlesticks, candelabra, and similar metal goods, and as this required a foundry, we were again obliged to move. Even then, in 1891, we had fully made up our minds to erect a suitable building for our work, but as that could not be done in a day,

we hired, for the time, the additional premises 38 Dey Street, where we occupied all the lofts.

In March, 1894, our new factory, De Kalb Avenue and Rockwell Place, Brooklyn, was finished. The illustration of the building will give some idea of its spaciousness. As it was designed and planned expressly for our use and is equipped with the latest improved machinery, it will be seen that we have all the facilities for manufacturing to advantage every article in our line. There are three separate departments in our factory, viz ·

THE GOLDSMITHS' AND SILVERSMITHS'.
ECCLESIASTICAL METAL WORK.
THE MANUFACTURING OF GOLD AND SILVER MEDALS.

In the hope of interesting our friends, we give in the following pages a description of the several kinds of arts and trades carried on in our Brooklyn works.

We take this opportunity to thank the many friends who have favored us with their orders in the past, and to bespeak for our home industry a continuance of their patronage. Our aim has always been, and ever will be, to give entire satisfaction in all we sell, both in price and in the quality, artistic and mechanical, of our work.

Designing and Modeling.

The first step in the manufacture of ornamental work of any description is a good design, and the great essential for a good design is a competent artist.

Come with us into our designing room, and peep over the shoulders of those engaged there. Here, apart from all disturbing elements and surrounded by works of art from which to drink in inspiration, each is busy at his specialty. This one, a man of creative genius, is absorbed on a first rough sketch destined finally to develop, under the cunning fingers of the skilled silversmith, into a beautiful ostensorium. Another is engaged on a finished drawing which exhibits carefully every little detail of an exquisite crozier. A third, less of the artist and more of the mathematician, with square and compass close at hand, gives his attention to what is known as a "working drawing," which represents every piece of the finished work in its actual size,

MODELING

from the boldest parts to the tiniest screw used in its con- . struction. Still another. with no guide but the flat drawing before him, its rounded parts distinguished from the others merely by the difference in light and shade. is modeling in wax the figure of an angel to be used on a candelabrum. How deftly he works the yielding mass, now with his fingers, again with a small tool, until beneath his magic touch the shapeless lump assumes symmetry and form!

In modeling, wax is used when delicacy and minuteness are required, as in the intricate and artistic forms peculiar to the work of the goldsmith and the silversmith. But for large models, which admit of more freedom in their manipulation, clay is employed.

Apart from the constant production of new designs for our regular stock goods. which is necessary if we would keep abreast with the times, much of the work on which our artists are engaged is for special orders. Our facilities for executing these are unlimited. and our success in this line is attested not only by many unsolicited testimonials. but by that unfailing sign of approval : repeated orders.

DESIGNING.

8

Gold and Silversmiths' Department.

And now if our friends will follow us we shall be happy to show them over our shops. and explain the many and varied processes carried on there.

Next to gold, silver is the finest and the most malleable of all metals. While harder than gold, yet in a pure state it is so soft that it can be easily cut with a knife. This makes it necessary. in manufacturing, to combine it with some other and harder metal which adds strength.

In casting the precious metals great care is taken that no part be wasted. For this reason a foundry is set apart for their use, and here by means of a gas furnace the gold and silver, alloyed with a small proportion of other metals, are moulded into ingots or cast into statuettes and delicate ornaments. As in the other departments, where the precious metals are worked, every tiny scraping is saved—even the sweepings of the floor, mixed though they are with dirt—and by refining brought back to its former pure state.

An important factor in the manufacture of all metal goods, and among the first machines to which we would call attention, are the Rolling Presses, into which thick strips of metal are fed. While their simplicity and efficiency render them liable to catch the eye of the visitor, let him beware lest they catch his finger, for such is their power, they would roll it out longer than the tradi-

11

tional finger of Time. After the removal of the bar of metal from the ingot mould it is ready for the operation of rolling. By means of readily-adjusted steel rollers, revolving one above the other, bars of metal are lengthened out and reduced, when necessary, to a thickness less than that of thin paper. This is not done all at once, but by passing the metal many times

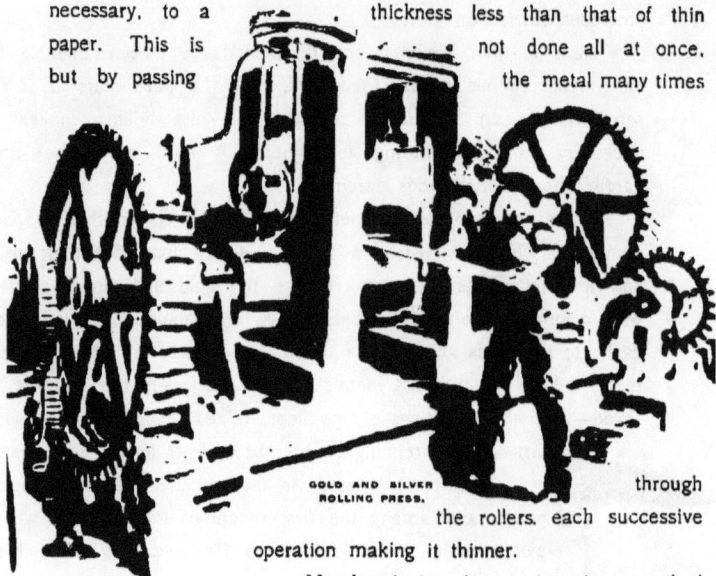

GOLD AND SILVER
ROLLING PRESS.

through the rollers, each successive operation making it thinner.

Metal spinning is an ingenious method by which a flat bit of metal is in a few moments fashioned into a cup or other piece of hollow ware without dies or casting, without striking a blow by hammer or press. As the manipulation is fully described on page 34, we will leave it for the present and turn our attention to the chaser, who is busily engaged in ornamenting the base of a chalice.

Chasing is an art that dates back to a very early period. Phidias and Polycletus. Greek sculptors who flourished before the Christian era, were celebrated for their skill in the art, and in modern times Benvenuto Cellini first attracted attention by his ability as a chaser and gold worker. The goldsmith's and silversmith's work of the sixteenth century reached its greatest splendor and beauty in his hands. He represents the goldsmiths and the silversmiths of the Renaissance. as Michael Angelo and Raphael represent the painters and sculptors. At the age of thirteen Cellini was apprenticed to Michael Angelo. From him he went to the workshops of many goldsmiths in Florence. Pisa, Bologna. and Siena. At the age of nineteen he went to Rome, and some years later entered the service of Pope Clement VII., and worked for him for fourteen years. Afterward he took service with Cosmo dei Medici in Florence, for whom he undertook the mint, made beautiful jewels, and

executed several important pieces of bronze sculpture. He covered the vessels he made with small figures, such as a chalice of gold ordered by Pope Clement VII., the cup of which was supported by the theological virtues. His jewels were enriched with figures on a minute scale. A necklace containing a history of the Passion, with separate compositions in each of its links, is still shown as an example of his genius. A book of hours, the cover of which is ornamented with little figures and compositions in enameled gold, is attributed to him. A salt cellar of his workmanship is in the Museum of Vienna, and many rich and costly cups and vases made by him are still preserved. He was a contemporary and admirer of the great Italian artists of his day, and his art represents the ideas then popular.

A CHUCK.

Many of the greatest painters. sculptors, and architects of the revival received their education in art in the workshops of master goldsmiths. Francesco Francia, a goldsmith of Bologna. is spoken of for the excellence of his enameling on metal in relief. He was also celebrated as a sinker or cutter of dies for coinage and medals. He did not learn to paint till after he had grown to manhood, though it is as a painter that he afterward became famous. His metal work, so far as we can judge of it from his

paintings, like that of Botticelli, also a designer of metal work, partook of the tender and serious beauty that belonged to the earlier times. Domenico Ghirlandajo, so called from the garlands he made of jewels for the Florentines, was trained under a goldsmith, though he is known to us by his paintings. A still more celebrated name is that of Andrea del Verrocchio, the master of Leonardo da Vinci in painting, and the sculptor of the statue of Bartolommeo Coleoni in front of the Church of SS. Giovanni è Paolo in Venice, the earliest and the grandest of modern equestrian statues. He was among the goldsmiths employed on the silver altar of St. John. He was sent for by Pope Sixtus IV. to restore the images of the Apostles in the pontifical chapel. Another goldsmith of great name was Ambrogio Foppa of Milan. He was skilled in the whole range of goldsmith's work, principally in enameling on relief and in medal cutting. Michelagnolo di Giuliana was a goldsmith of Florence, much employed by Lorenzo and Giuliano de Medici, for whom he embossed armor, enamels, and jewelry of every kind. He was the first teacher of Benvenuto Cellini. We trust our readers will pardon this digression, which is merely to show the artistic standing of our art.

Chasing is of two kinds: the one in which the artist creates by both brain and tool; the other in which he merely copies the design before him. In both instances, however, an artist is necessary, it being a question of excellence only, for the delicate

touch which executes the finer kinds of chasing must, in all cases, be directed not only by skill and knowledge, but by that sort of culture known as "the artist instinct."

Here is a chaser of unusual excellence. Let us watch him at his work. Taking the base of the chalice, he by gentle but repeated blows of his "snarling iron" on the inner side of the metal forces certain parts into relief, so that they stand much higher than the surface. This being done to his satisfaction, he fills the inside of the base with a composition of molten pitch dipped from the immense caldron in a steam oven where it is kept hot. The mixture soon cools, and becoming solid forms a foundation which preserves the parts in relief. Now begins the actual work of chasing. The metal is not cut away, as in engraving, but by light taps of a hammer upon a simple tool, the chaser drives in the metal, and indents it until it assumes the desired pattern. Chasing is, necessarily, a slow operation, weeks of unremitting labor being frequently expended on a single piece of work, but when finished the exquisite beauty attained fully compensates for the time devoted to it. Many of the art treasures of the world consist of exquisite

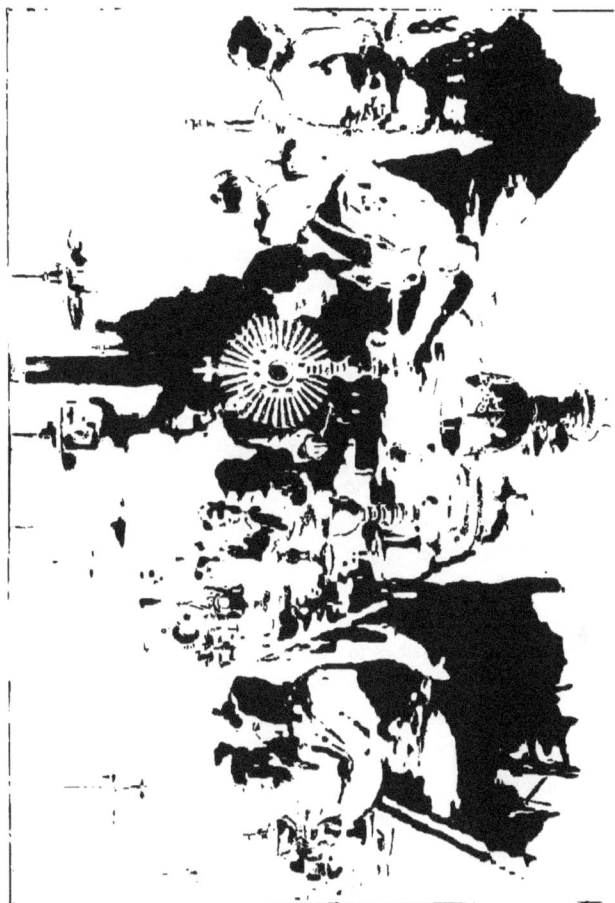

SILVERSMITHS.

examples of the chaser's art, their value depending many more times on the labor expended than on the metal of which they are made. Most of the art treasures in silver and gold which France possessed, however, were destroyed after the death of Louis XVI. The greater part of the ancient shrines, chalices, reliquaries, croziers, and other sacred utensils were seized by the commissioners and sent to the revolutionary mint. This was also the case in other countries. In Italy, Spain, and Malta, wherever the French armies were in possession, all which could not be removed or hidden was seized and sent to Paris. In 1810 the French sent a commission to the Escurial, who took possession of the treasures there, only allowing the friars to remove the relics from the reliquaries. They broke the caskets and jewels and threw the relics to the friars. It is impossible to describe the wanton destruction and robbery committed in the Spanish churches, where was destroyed the largest collection of art objects of gold and silver workmanship in Europe. From the Cathedral of Leon alone more than 10,000 lbs. weight of old silver was carried away.

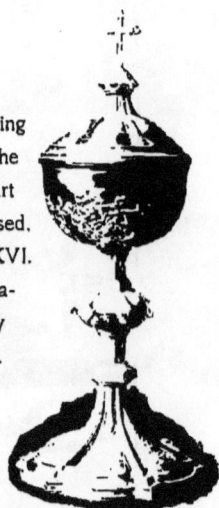

Some rich specimens of the art, in solid silver and in gold, have been executed in this department of our factory.

In engraving, the metal is cut with a sharp tool, and this also calls for a high degree of skill and taste.

Leaving the chaser and engraver, let us turn

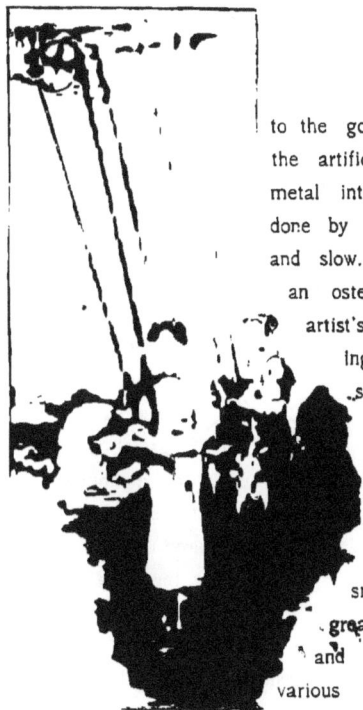

to the goldsmith and silversmith proper. He is the artificer who with his hammer fashions the metal into the required shape. Like all work done by hand, that of the silversmith is difficult and slow. See this one making the rays for an ostensorium. Before him is spread the artist's design. This he traces on the shining metal, then begins to beat it into shape, and finally with hammer and chisel cuts the rays to the required form. Then he solders them to a ring. But don't confound this process with that of the tinsmith or the itinerant tinker. Soldering as applied to silversmiths' work is an art which requires great care and practice to perform neatly and properly. It consists in uniting the various pieces of an article together at their junctions, edges, or surfaces by means of an alloy, specially prepared for the purpose, which is more fusible than the metal to be soldered. The solder must in every way be suited to the particular metal to which it is to be applied, and must possess a chemical affinity for it; if this be not the case, strong, clean, and invisible connections cannot be effected, whilst the progress of the work would be considerably retarded. The best connections are

made when the metal and the solder agree as nearly as possible
in uniformity as regards fusibility, hardness, and malleability. Our
solder is an alloy of silver. It is applied in a powdered state to
the articles to be united; they are then placed on a furnace,

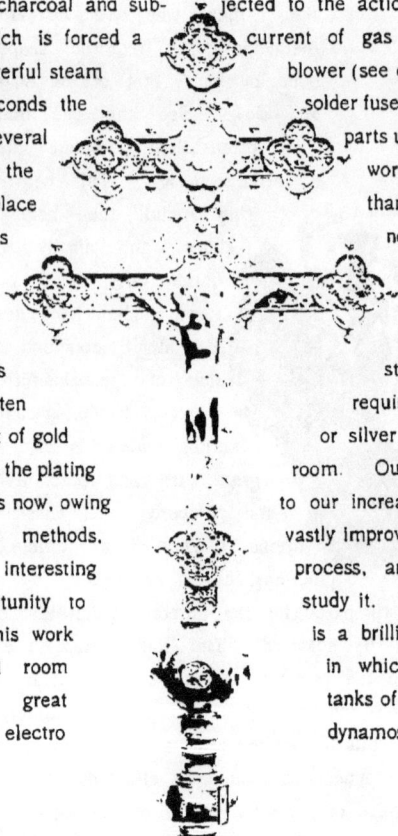

covered with charcoal and sub- jected to the action of a blow-
pipe from which is forced a current of gas mixed with
air from a powerful steam blower (see cut. page 34).
In a few seconds the solder fuses, and running
between the several parts unites them so
thoroughly that the work will break in
any other place than on the joints.
The work is now cleaned by
" buffing," described at
length on page 38,
and is ready for mounting
with precious stones, if necessary, as is often required.
Goods not of gold or silver, that have to
be plated, go to the plating room. Our plated ware,
always good, is now, owing to our increased facilities
and superior methods, vastly improved. Electro-
plating is an interesting process, and here we
have an opportunity to study it. Set apart especially for this work is a brilliantly lighted,
well ventilated room in which are huge
earthen jars, great tanks of water and of
chemicals, and electro dynamos. The fumes

PLATING ROOM.

19

of acids and the gases common to such a room are here,
by proper apparatus, carried to the open air to be dis-
sipated. The article to be plated is thoroughly cleaned,
as any surface-impurity would spoil the success of the
operation. It is first boiled in caustic potash to remove all grease, then immersed in acid to remove
any rust or oxide that may have formed, and lastly
scoured with fine sand. It is now ready for the
plating-bath of gold or silver, as the case may be.
By means of a wire from a metal rod lying across
the jar, the article to be plated is immersed in the solution. A plate of pure metal, termed an anode, is suspended
in the bath by a second wire; the electric current is turned
on and the pure metal is deposited on the suspended
article. The thickness of the plate
depends on the length of time the operation
is continued.

An experienced electro-plater can produce
various shades of gold, such as light, dark,
yellow, red, green, etc., on the one article, and
the contrast is pleasing and beautiful.

When the plated object is taken from, say,
the silver bath, it appears dull and white, but its
appearance is soon changed by scratch-brushing, un-
less required to be left a *dead* white, when this
process does not take place. Scratch-brushing
is one of the indispensable and constantly re-

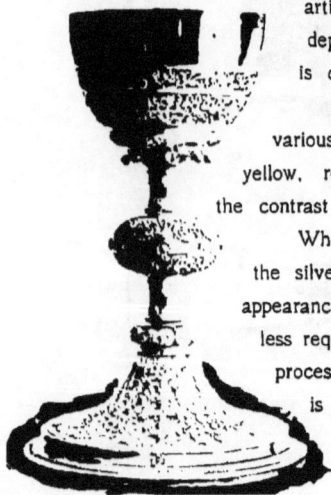

curring operations of the electro-plater's art, and nearly always follows the plating, The instrument used for this purpose is called a *scratch-brush*, and its shape varies with the article it is to be used upon. Scratch-brushing is seldom done dry; the tool as well as the article operated on must be kept constantly wet with a solution which may set up a chemical action, but which generally acts merely as a lubricant, and at the same time carries away the impurities that the brush detaches. The scratching removes the dull, white color from the surface of the metal and gives a characteristic brightness to the work of the silversmith. Scratch-brushing, while sometimes operated by hand, is generally done by a fine brass-wire brush of circular form running upon a spindle and driven by a lathe.

Burnishing is the next operation. It produces a polished surface, which reflects like a mirror, and gives the highest lustre; it removes marks left by the polishing mixtures, and produces a darker surface than the other modes of finishing. The tools employed for this process are extremely variable, and well adapted to the different kinds of work to which they are applied, but all must fulfill the requisite conditions of great hardness and a perfect polish; they are of two kinds, one being of hard stone and the other of polished

steel; they vary in shape, some being straight with rounded points or with curved and blunt edges, others with large rounded surfaces. The stone burnishers are made of blood-stone, mounted in a wooden handle. For very small articles steel burnishers only are used, as they are of a finer make and by their greater variety of form are well adapted to all kinds of work. Burnishing may be applied by either hand or machine. A large, plain object, as, for instance, the cup of a chalice, is placed on a wooden or metal form fastened to a lathe, and while this is made rapidly to revolve, the workman applies to its surface a burnisher which soon communicates a brilliant polish to the cup. The burnisher's tool and the article being burnished are frequently moistened with certain solutions, some of which merely facilitate the sliding of the tool while others have a chemical action which effects the shade of the burnished article. Smaller and more delicate objects are treated by hand; no lathe is used, but careful and patient rubbing takes its place. When the burnishing is completed the surface of the article is wiped with an old, soft, cotton rag; sawdust, hard cloth, and tissue paper produce streaks. The finish obtained by burnishing is satisfactory when the article reflects the luminous rays like a mirror. An experienced workman can produce wonderful effects of light and shade by skillful management of his burnisher.

Finally, the several parts are put together, the whole is carefully scrutinized, and if found right the article is finished.

Ecclesiastical Metal Work.

In this department are manufactured Candlesticks, Crucifixes, Candelabra, Sanctuary Lamps, Processional Crosses, Altar-Rails, Pulpits, Memorial Tablets, and the hundred and one other Church goods which are made either to special order or for stock.

Not many years ago it was thought impossible to make goods of this kind in this country which would compete in price and quality with the imported article. But there is no longer any question of it, for the products of our factory, while quite as cheap as the foreign articles, are fully equal in style, workmanship, and durability.

Of all the arts there is perhaps none so ancient as that of the artificer in brass, for from the earliest time every country has had its brass workers. Nearly six thousand years ago the metallic strata which cropped out from the mountain sides were fired, hammered, and rendered submissive to the will of the great father of metal workers, Tubal Cain. This extraordinary man gathered around him a school of artificers, and unceasingly labored to make them acquainted with the properties of brass, the mode of its reduction, and its application to his many inven-   tions.

In early Hebrew writings we have such expressions as " mountains of brass," " brass molten from stone,"

A PATTERN.

from which some infer that the ancients obtained it in a native form. We are familiar with the word brass in its frequent application to statuary, bells, gates, and the like. It was a brazen serpent which Moses raised in the wilderness, Goliath "had a helmet of brass upon his head," and the swords, armor, and cooking utensils exhumed at Ninive are of brass.

Heathen mythology, too, favored the brasier's art. Brass or bronze statues were made in great quantities, and on them often depended the wealth of a state. Athens and Rhodes had at least three thousand statues each. The famous Colossus of the latter place, erected

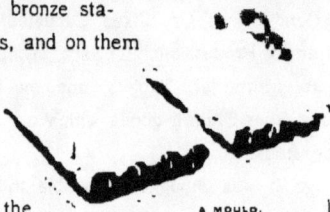

A MOULD.

to Apollo three hundred years before Christ, measured 105 feet in height, took twelve years to execute, and cost the enormous sum of $590,000. The statue was destroyed by an earthquake fifty-six years after its erection, and remained buried for over nine hundred years, when it was sold by the Saracens to a Jew, who extricated from it nine hundred camel loads of brass. During the early ages of Christianity we find brass bells introduced into the church towers, to call the people to divine worship. The earliest traces of brass in Great Britain are found in the mediaeval monumental brasses over the tombs of civil and ecclesiastical dignitaries. These "brasses" were somewhat similar to the memorial tablets of to-day, with the exception that

24

FOUNDRY
IMPLEMENTS.

they were generally let into the pavement above a tomb. They were not confined to England, but were common in France, Germany, Flanders, and the Netherlands. Unfortunately, the intrinsic value of the metal led to the wholesale spoliation of these interesting monuments. In France, those that escaped the troubles of the sixteenth century were swept away during the Reign of Terror. The fine

CASTING.

memorials of the royal house of Saxony in the cathedrals of Weissen and Freiburg are the most artistic and striking brasses in Germany. Among the thirteenth century examples existing in Germany may be mentioned the full length memorials of Yso Von Welpe, bishop of Verden (1231), and of Bernard, bishop of Paderborn (1340). Many fine Flemish specimens still exist in Belgium.

"Figure brasses" of Flemish origin are found both at Bruges and in England. They have the figures engraved in the centre of a large plate, the background filled in with diapered or scroll work, and the inscriptions placed around the edge of the plate.

Having now, we trust, duly impressed our readers with the importance of brass work-ing, let us pass to details.

The very corner-stone of this metal work department is the FOUNDRY. Leaving the first floor with its hum and whirl of machinery, we enter a large, well ventilated room. across which grimy figures flit silenty to and fro. To the right, under a vast skylight, men are busy about a table filled with what appears to be a rich, black earth, but is, in fact, "moulding sand." Pick-ing up a box composed of two open iron frames, the one fitting exactly on the other by means of dowels, and technically known as a "flask," the moulder places it on a flat board and proceeds to fill one frame with sand; in this he places a pattern of the article to be cast. The other frame is then put on, more sand is added until the pattern is well covered, and finally the moulder shovels in moist sand and rams it tight with a wooden rammer, till the flask can hold no more. It is then scraped flush, and pricked with a pointed wire, to give vent to the gas formed in pouring in the metal. A second flat

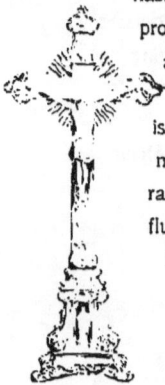

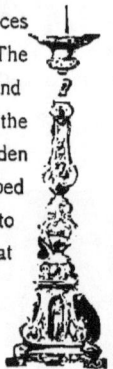

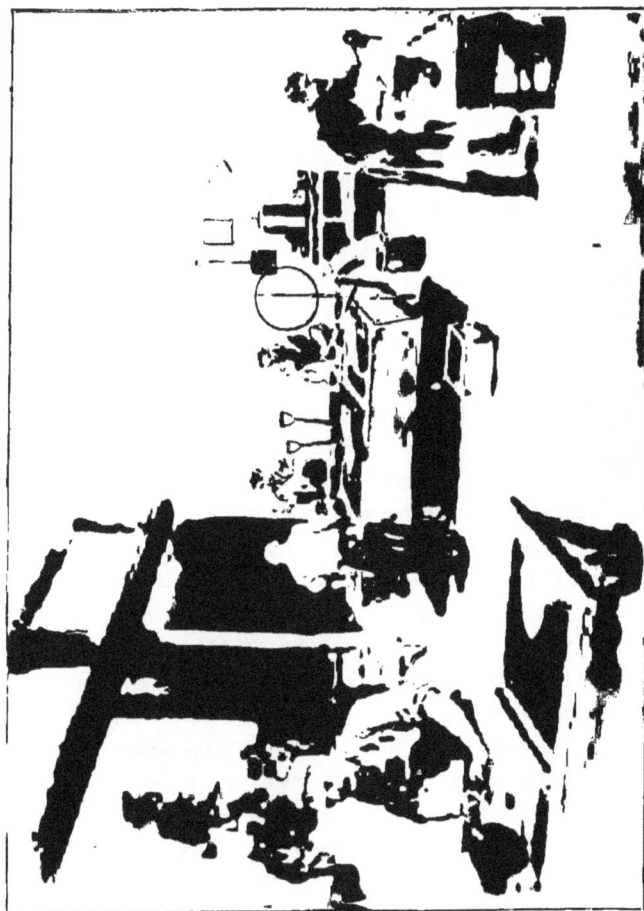

FOUNDRY.

27

board is placed on the top like a lid, and both boards and flask are carefully turned over: what was the bottom part is then removed, showing the pattern imbedded in the sand. The operation of withdrawing the pattern is done by slowly tapping it on all sides to relieve it from the sand, and then with a steady hand lifting it

A DIE.

out. If any of the sand be disturbed in doing this, the mould must be repaired; and any of the intended castings which are to have holes or hollows are supplied with pieces of well-baked sand called cores. The moulder having fixed these, he cuts passages, or channels of

DROP PRESS.

communication, between the pattern and the exterior of the flask for the admission of the molten metal. In ramming the flask some art and practice are required. If rammed too hard the generated gases do not escape freely, and air holes in the castings are the result; if too soft, the pressure of the metal causes swellings on the castings. In either case they are useless. The mould being perfect, it is placed in position for the next step, that of casting. On the left side of the entrance, over against the wall, are vast furnaces sunk in the ground, their tops on a level with the floor. Down in these fiery caverns are the crucibles of molten brass. Grasping an enormous tongs, a sturdy fellow throws open the furnace top, and reaches over

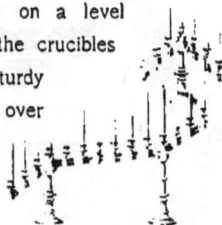

28

BENZIGER BROTHERS' SINGLE ROW
PATENT CANDELABRAS.

to seize a crucible. Standing thus over this sea of fire, his face and form lighted by the lambent flames, we are reminded of Vulcan in his palace on Olympus "brazen, shining like the stars." Lifting the pot of liquid metal to the surface and carrying it to the mould, the man (see page 25) tilts it over and pours the metal into the mould through holes, called *gates*, until it flows over, running like water on the brick floor. The work is done; the crucible is returned to the furnace, rough casting taken out and thrown into a tub of water to cool.

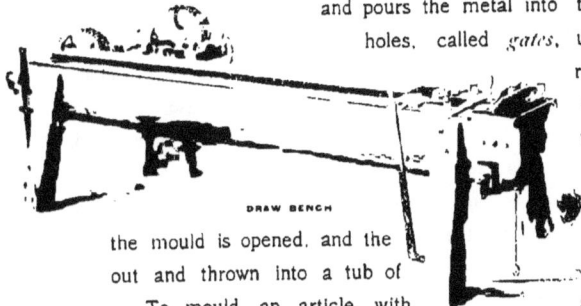

DRAW BENCH

the mould is opened, and the

To mould an article with a smooth surface is a very easy performance, but it is different with more complicated forms, such, for instance, as a statue or any piece in which certain parts are in high relief while other parts are greatly depressed. For such articles resort is had to what are termed *false cores*. A core is that part of a mould which shapes the inside of a hollow casting, or which makes a hole through a casting. A false core is a piece built up separately from the mould proper, and can be displaced to allow the removal of the pattern. Suppose a small statue of Our Lord is to be moulded; while it is easy enough to bury the pattern in the sand of the mould, it is impossible to withdraw it without breaking the mould on account of the depressed parts, such as under the chin, the space between

DRAW PLATE.

the arms and the body, where the sleeves of the gar-
ment fall about the wrists. etc. To overcome this difficulty.
only half of the statue pattern is moulded at first; then
the surface of the mould is polished, so as to be perfectly
smooth, and the other parts are built up of as many separate pieces as are necessary. The sand of these pieces is
packed tightly by being pressed into place and hammered,
and when it has set. each piece can be removed at pleasure,
and there is no longer any difficulty in taking   the statue
pattern out of. the mould. When that   is removed the false cores are fastened in
their places by pins. and the mould is
ready for the metal. Sometimes a
dozen, or even more, of these false
cores are necessary.

The original *patterns* from which
castings are formed are made of
wood. smoothed and varnished; but
for goods that are to be reproduced
in quantities the patterns are
moulded in brass, and carefully
finished by the chaser. Sometimes, too, the original pattern
is of brass polished and finished
with as much care as are the
goods themselves.

As will be readily understood,

30

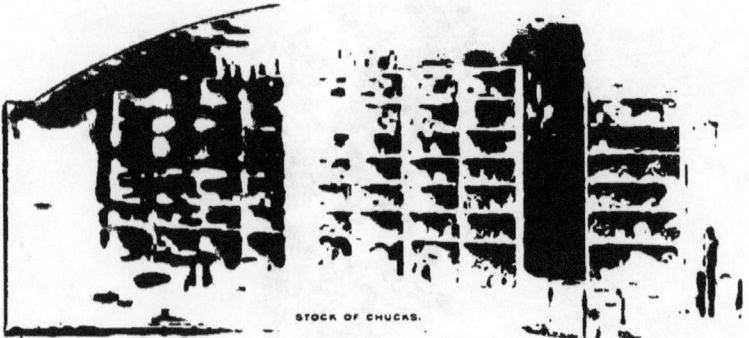

STOCK OF CHUCKS.

these patterns are of great value to the manu-
facturer, and as they represent a considerable outlay of time and
money they are kept, each with its distinguishing number, in a
spacious fire-proof vault adjoining the foundry.

Beside the patterns, the dies are also kept here. There is a
great difference between a pattern and a die. The one is in
relief, the other is sunk. The one is used for casting, the
other for striking up sheet metal, making medals, etc. A
die is hollowed out of solid steel. It is the very reverse
of the chaser's work, and yet produces a somewhat
similar effect, with this advantage, that, when finished,
it can by a single blow and in a moment's time
do that which would cost a chaser hours, sometimes weeks, of labor. Most of the smaller and
finer ornaments on our goods are made by dies, and
they are also used to give the finishing touch to
small castings.

The die is worked by a press. For the heavier patterns
on thick metal a DROP PRESS is used (see page 28); this

is a primitive sort of affair: being merely a heavy block of iron which falls from a height and by its weight drives the die into the metal. Though simple, it is wonderfully effective, and there is little prospect that it will be superseded by any more elaborate machine.

The machine which next attracts our attention is a DRAW BENCH (see page 29), used for making wire. An oblong

CHASER.

PATTERNMAKER AT START.

PATTERNMAKER AT FINISH.

plate of hardened steel is set up on edge. This plate is pierced with a number of conical holes, each of a different size, and each gradually diminishing in diameter until the smaller end is the size of the wire required. A strip of metal, pointed at one end, is placed through the largest hole, whereupon a clamp seizes it, and with irresistible power draws it through one hole of the plate after another until it is reduced to the required size. The clamp is automatic and returns of itself to seize the wire afresh when it has drawn it to the end of the bench.

33

We have now reached one of the most interesting operations in our factory, that of METAL SPINNING. The requisites for this work are a lathe, certain wood or metal forms, technically known as *chucks*, a blunt steel tool, and, *facile princeps*, an intelligent workman. A chuck of the required form being fastened to the lathe, the workman places on it a disk of metal: gold or silver,

SOLDERING.

copper or brass, and starts the lathe. As it swiftly revolves he applies the tool and gradually bends the yielding metal against the chuck until it assumes every line and curve of the form. In this way a solid cup may be, and is, formed without seam of any kind, and yet, as we have before stated, without die or casting, without striking a blow. As there is no slight pressure brought to bear upon the chucks, they have to be made of very

hard wood, and dogwood is chosen as best
suited. But all wood being more or less porous,
the metal is apt, at times, to show a certain
roughness where it has sunk into the pores, and
for that reason, as well as for their greater
durability, iron chucks are sometimes used,
especially for articles that are made in
large quantities.

The man we see at the other lathe
is TURNING. His work is of such a
multifarious character that a volume might
be devoted to it. He has to make all kinds of
plain, spherical, and ornamental turning, which necessitate
a complication of lathes and tools. In the turner's hand
the tool must never be allowed to rest, but must have
a proper rotation and a correct inclination to avoid
furrowing the work. It is the turner who makes the
"thread" of a screw; bores long
holes through solid rods of steel,
iron, or brass; and converts un-
shapely pieces of metal into forms
of use if not exactly of beauty.

The work of chasing has already
been described, but here are several
chasers (see page 33) engaged in
pattern making, working up solid
metal into beautiful and fanciful

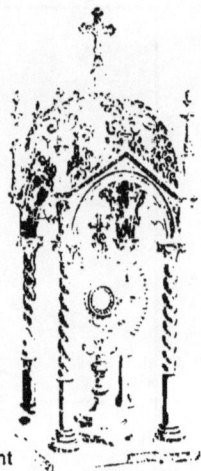

·6

forms, which the foundry will reproduce
many, many times by casting.

As will be seen, it requires men of
many trades to complete what appears to be
a very simple article. The casting as it
drops from its mould in the foundry must
pass through many hands before it becomes
the beautiful, highly-polished candelabrum or
crucifix. The die-sinker, the chaser, the
turner, the burnisher, must each help
towards its completion. Take this small
figure of an angel, as an example : at first
it is in several parts, and rough parts at
that, but this man files away the adhering,
superfluous metal, while that one adjusts the wings and solders
them in place, and so it passes from one
to another until it stands finished in form
and feature, a thing of art to delight the eye.

One of the last operations before put-
ting an article together is the BUFFING. The
buffing or polishing machines in our factory
occupy the half of one floor. The article
to be polished is first cleaned by swiftly-
revolving steel brushes which remove all
foreign substances and brighten the metal.
It is then dipped in acid and in lye, to
further clean it, and finally is ready for

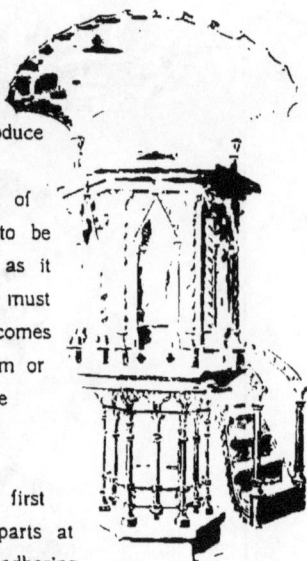

LACQUERING ROOM.

polishing. This is done by means of a cotton wheel covered with *rouge*. The wheel is attached to a lathe and revolves with a high degree of velocity. The dust and dirt generated by this machine are collected and drawn into pipes by a blower or "exhaust draught," as it is called, which carries the waste into large tanks of water, and thus keeps the floor free of the shavings and dirt thrown off by the buff.

Not all that glitters is gold, and not all our brass work is gilt. Much of it owes its brilliancy and color to thorough burnishing or buffing and to lacquering. An article of brass that is highly polished greatly resembles gold. To heighten this resemblance and, in a measure, to prevent tarnishing, lacquering is resorted to. The polished goods are picked up very gingerly, so as not to impair their lustre, and treated to a coat of lacquer, a sort of varnish, of the required color. They are then placed in a drying oven, whence they are removed ready for mounting.

MOUNTING is the last step and one that requires a true eye, a steady hand, and good judgment.

39

BENZIGER BROTHERS' DOUBLE ROW
PATENT CANDELABRAS

DRYING OVEN.

Look at these men "mounting" an altar-rail. How closely they scrutinize each piece to see that it has its proper finish; how carefully they put the parts together, testing the screws, turning them back and forth, to assure themselves that all run smoothly; and, finally, fitting the separate parts one to another until the entire rail stands completed, just as it will appear when in its destined place in the church.

Just here we might add a few words in regard to the great growth of the metal altar-rail. Until within a few years such a thing was hardly known. Only the heavy, architecturally severe wooden rail was to be seen in our churches, but now, thanks to the ever-increasing appreciation of the beautiful which marks our age, this unsightly object has given way to the lighter metal rail, made in a variety of graceful and artistic designs.

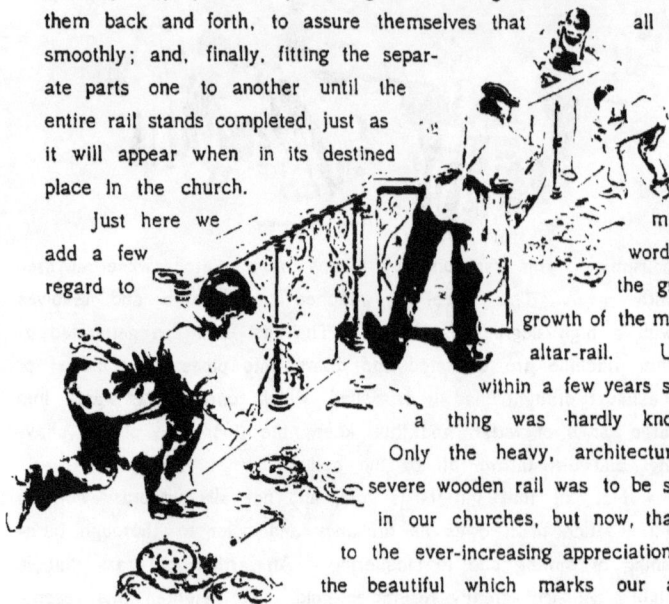

SETTING AN ALTAR-RAIL

We are devoting special attention to the manufacture of these metal altar-rails, and among the many patterns, of our own exclusive design, are some of rare grace and beauty.

Our Medal Department.

The manufacture of gold and silver medals has, within a few years, grown to be a feature of our business, and we are no longer obliged as formerly to go abroad for our supply. Medals are of two periods. ancient and modern. To the former belong those issued from the mint of ancient Rome, to commemorate some notable event in the history of the nation. and the beautiful medals of the ancient Greeks and Sicilians. Modern medals begin with the fourteenth century, but most of them were not struck until a century later. Most European nations possess a succession of medals from the fifteenth century onwards. The best in point of design of the fifteenth century medals are those made by Victor Pisani. The medals of the Popes form an unbroken series from the time of Pope Paul II.. who occupied the throne of Peter from 1464 to 1471 ; their reverse generally bears the cross-keys and mitre, while on the obverse is the head of the reigning Pope. Some of the medals of Julius II.. Leo X.. and Clement VII. have an especial value as having been designed by Raphael and Giulio Romano and engraved by Benvenuto Cellini. France produced

EMBOSSING PRESS.

few medals prior to the time of Louis XIV. The Spanish medals begin with Gonsalvo, about the year 1500.
Scot land produced one of the earliest of modern medals,
struck by David II. English medals begin with Henry VIII.
Here, in America, we have a series of beautiful bronze
medals struck by the Philadelphia mint, each bearing
the head of one of our Presidents, from the time of
Washington down to the present day. The Church
commends the wearing of medals bearing religious
emblems as badges, to show the faith and fealty of her
children. Preëmi- nently the most popular of all
medals is the Miracu- lous Medal of the Blessed Vir-
gin. Though there are many other medals approved
and blessed by the Church, there is none to compare
with this little symbol of our confidence in our
Immaculate Mother.

The manufacture of medals calls for the ser-
vices of an accomplished die-sinker and several
elaborate machines. The metal is first brought to the
required thickness by the rolling press, then passed

through steel roller-dies which em- boss on
it the outlines of the medal, cut into shape
by another press, and finally stamped with
the full impression. Soldering on the
rings, cleaning the medals and gathering
A DIE. them into dozens, are still further operations
necessary before they can be sent to our salesrooms.

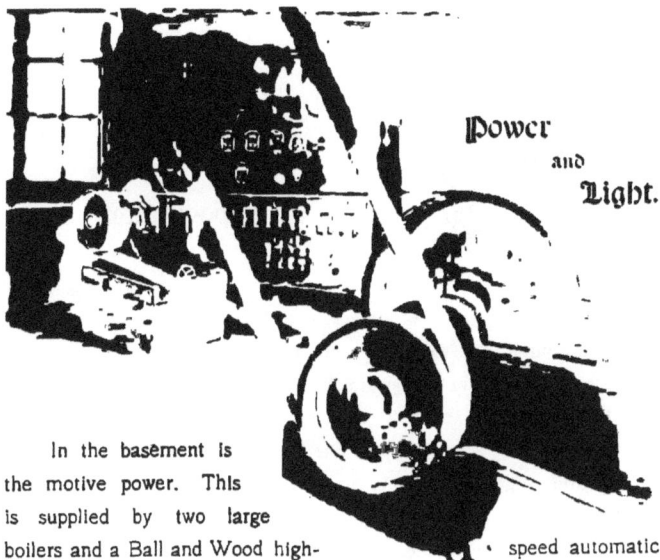

Power and Light.

In the basement is the motive power. This is supplied by two large boilers and a Ball and Wood high-speed automatic engine. This engine drives a dynamo of 55 kilo watts. equal to 74 horse power. and this, in turn. supplies the different motors in the building. Electricity as a motive power, though a new departure, is in every way practicable, for it not only dispenses, to a great extent, with shafts and belting, but does away with the necessity of cutting the floors through which to run them. The motors vary in horse power, and as each group of machinery has its own motor, one may be shut off, when not in use, without stopping the whole of the running gear. Electricity is also used to run our freight elevator, and, what is still more important, to light the factory. To the latter purpose 300 incandescent lamps. each of 16 candle power. contribute, beside a large arc light used in the foundry. The same power drives the " exhaust " which carries off the dust and dirt.

Expositions.

In the VATICAN EXPOSITION of 1887, with the whole world of Church goods manufacturers represented, we were not only accorded THE DIPLOMA OF HONOR, the highest kind of award bestowed, which carries at the same time the right to the gold medal, but we afterwards received a still further proof of the merits of our manufactures, and to us a still higher honor, in the conferring on us by HIS HOLINESS POPE LEO XIII. of the title

"PONTIFICAL INSTITUTE OF CHRISTIAN ART,"

with the right to the use of the Papal Arms.

Visitors to the COLUMBIAN EXPOSITION of 1893, in Chicago, may recall our pavilion, near the centre of the large building of Manufactures and Liberal Arts, in which we had a rich display of the goods made in our factory. For this exhibit we received the highest award given at the Exposition, viz: the diploma, and the award for "High-class workmanship in the production of gold and silver and plated Church ware, and in the adherence in their production to the true ecclesiastical style." Our pavilion occupied a prominent position in the

ESPOSIZIONE VATICANA

LEONE XIII

44

main transverse aisle. The large octagonal case contained some very remarkable examples of the work done in our Silversmiths' Department, ranging from plain chalices, ciboriums, etc., to the finest specimens of the art finished in rich repoussé work. In the centre stood a Gothic ostensorium, 40 inches in height, designed and made expressly for the Exposition, an illustration of which is on page 11. The great objection to such a large size lies in the fact that it is generally too heavy for practical use. This was overcome, however, and the ostensorium rendered comparatively light, by making the columns which support the baldachins hollow, and making them by hand instead of casting them. The large case at the rear exhibited specimens of the skill of our Metal Work Department, from candlesticks and candelabra of the smallest size, intended for chapels, to the very largest kinds suitable for cathedrals.

BENZIGER BROTHERS' EXHIBIT AT THE WORLD'S COLUMBIAN EXPOSITION, CHICAGO. 1893.

Our Salesrooms and our Catalogues.

A visit to our salesrooms will give a still clearer idea of the facilities we possess for manufacturing everything connected with our business, for there will be found in endless variety a complete stock of Church goods of every description.

The New York house, founded in 1853, occupies the entire building Nos. 36 and 38 Barclay Street; the Cincinnati house, established in 1860, is at 143 Main Street, and the Chicago house, the youngest, dating back only to 1887, is at 178 Monroe Street.

Besides the goods already enumerated, we carry a complete assortment of Regalia, made in our own embroidery department. Statues, Stations of the Cross, Stained-Glass Windows, and the Rosaries, Crucifixes, and sundry other things which come under the head of " Religious Articles "

In addition to our own publications, which embrace miscellaneous books, school-books, and prayer-

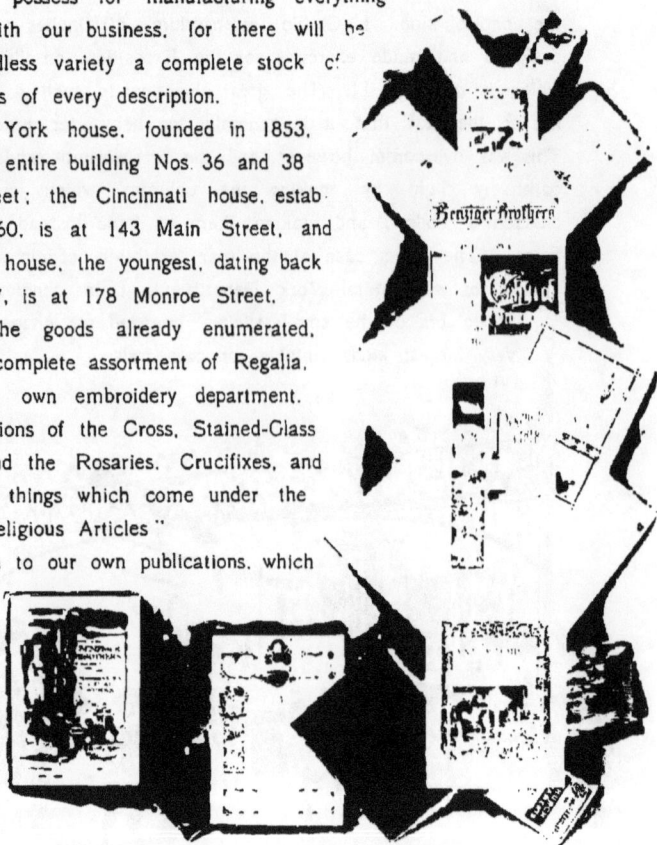

books, we keep a full line of the publications of other Catholic houses, American and foreign, of which we shall always be glad to furnish our customers with catalogues. Though our traveling agents visit our customers regularly, they must, in the selection of goods, rely in a great measure on our Catalogues. We therefore call especial attention to the Catalogue of our manufactures, which represents fully not only everything new in the service of the Church, in the many styles, but

gives a clear notion of the great number of patterns always in stock.

Our first illustrated Catalogue, published about 1864, when we began the business, contains about 50 articles of our own make. Since then our illustrated Catalogues, issued at regular intervals, have gradually increased in size, new goods being added and old ones dropped, until in the Complete Catalogue of manufactures of 1892 there appear no less than 898 illustrations.

www.ingramcontent.com/pod-product-compliance
Lightning Source LLC
Chambersburg PA
CBHW022029190326
41519CB00010B/1634